INSTRUCTIONS FOR DEALING WITH
SCHIZOPHRENIA: A SELF-HELP MANUAL

JACK K. BRAGEN

Copyright, December 2012, by Jack Kenneth
Bragen

ISBN: 978-1-300-38269-0

Questions? Contact the author at
bragenkjack@yahoo.com

DISCLAIMER:

The author of this book is a consumer, not a physician, and as such, can not disseminate medical advice. Please do not contact the author for any kind of advice, medical or practical.

This book shares ideas that work for me; these ideas may or may not work for the individual reader.

The author of this book is not liable for damages or for any mishaps resulting from reading this text, or from following the advice contained in it. All advice or ideas in this book are to be taken at your own risk. The author assumes that the reader has the necessary discretion to use common sense and to apply these ideas only where they are applicable. These ideas will not work in all situations, and it is up to the reader to decide whether or not to use them. Protecting yourself is up to you, the reader, and you are ultimately responsible for and accountable for your own actions.

If you do not understand the paragraph above, you should seek medical attention from a doctor or go to the nearest hospital emergency facility.

TABLE OF CONTENTS:

SPECIAL PRE-ADDENDUM: HOW TO DEAL WITH DELUSIONS

Copyright December 2012 by Jack K. Bragen

A delusion is a false and usually strange belief brought about by being psychotic. It is extremely important that delusions are kept in check, and are not acted upon. Delusional beliefs can be extremely dangerous when they are in control of your behavior.

If and when you are severely delusional, you must usually receive external intervention—the illness is in charge and it wipes out the judgment that would tell you that you need to get help. When severely delusional, you have lost the capacity to control your thoughts, and you actions will be based upon false beliefs.

Once medicated and stabilized, you will probably have some residual symptoms.

A delusion is an erroneous belief or thought which is created by psychosis. Psychosis is a state of delusions and/or hallucinations brought about by schizophrenic illness or by some other cause. A person can become psychotic from drugs. A person can also become psychotic by being deprived of sleep. A person probably has schizophrenia if they are persistently psychotic, and other causes of the psychosis have been ruled out.

Antipsychotic medication, when effective, alleviates some or most of the psychosis, and it helps reduce the quantity and the amplitude of delusions.

You can believe something untrue without being schizophrenic. The difference is partly that the schizophrenic person will often disregard their five senses in favor of the incorrect belief, and that person's thoughts and behavior will be disorganized. Secondly, a person with schizophrenia will have incorrect beliefs that aren't shared by other people.

It is probably next to impossible for someone to be completely without false beliefs, schizophrenic or not. However, when a false belief is revealed, the "normal" person (also, the person stabilized on medication) adjusts accordingly and will usually replace the false belief with the correct one. When you are psychotic, you probably can not do that.

Learning not to act on delusions is an acquired skill, and an important one. In part, it is a matter of learning to operate partly on an instinctive, gut feeling, and to sometimes prioritize this above what your thoughts are telling you. In part, it involves a learning curve, in which you use memory of a bad result when a similar action to the one you are contemplating was taken in the past. Learning not to act on delusions involves an extra bit of caution in which you use uncertainty to your advantage. If you are not sure about whether or not to do something, you probably shouldn't. You can be absolutely certain that you must get out of the way of a speeding bus. However, you should value your uncertainty when you have a thought of sending a letter to an elected official.

Learning not to act on delusions means that part of you is not delusional and recognizes the fact that you have false beliefs. Thus, accordingly, all of your actions employ extra caution.

The next step on the path of wellness is that you learn to deprogram you delusions. This can be done by evaluating a thought, and weighing it against your past "normal" experience. When evaluating a thought you should compare it to what was real in your past before you started to become ill. Or, you can compare your thought to what other people believe. When a thought seems unusual, if it seems strange, if it makes you different from other people, or if it makes you special in some way, you should question that thought.

Once you have pinpointed a thought that seems to be a delusion, you should tell yourself that it is a delusion and it is unreal. Your mind will understand this and will produce that thought less often or at lower amplitude. Each specific thought or sometimes a category of thought requires its own evaluation. It won't work to simply think, "I won't be delusional any more."

Pinpointing, not acting upon, and deprogramming delusions are tasks that should be done in addition to taking medication. Medication gets you into a range of brain function in which thinking normally becomes possible. Educating yourself on how to deal with delusions then becomes possible, and this in turn can sometimes give you a "normal" state of mind in which most of the thoughts are accurate.

FOREWORD

This book is dedicated to and written for those brave individuals who live every day in a battle against mental illness. We are people who try to live worthwhile existences. And our lot is more difficult than what most un-afflicted people could imagine.

This book is written by a man who suffers from mental illness, who has gone through most of the experiences that are common to people with mental illness, and who has gotten well enough to become a semiprofessional freelance author.

I have gone sixteen years since my most recent episode of psychosis that resulted in hospitalization. I continue to remain stable, on medications, and hope to never have another episode of acute schizophrenia. This book will describe how I accomplished sixteen years without a relapse.

I live to an extent as "normal people" do, except that I do not earn much money. I must rely on public benefits and on assistance from family, however, other than that, I handle my own affairs. This book will discuss the maintenance of one's own needs.

I have been a columnist for the Berkeley Daily Planet for the past one and a half years, and elsewhere I have written numerous freelance articles and short stories. I find that writing suits me partly because I don't have to leave the house to do it.

It takes bravery. It takes perseverance. And it takes effort. These are three traits that can get someone through the hard years of the illness, and will allow them to come out the other side with most faculties intact, hopefully to live a better life.

INTRODUCTION:

The life of a person with mental illness is hard.

The suffering of mental illness, (such as in a psychotic, manic or depressive episode) to begin with, is a very hard thing to go through. A psychotic episode can be terrifying, debilitating and immensely painful. It involves being cut off from having a rational mind. It involves being immersed in a false reality of imagined events that often make no sense and that make a person's consciousness disconnect from the external environment. The person is delusional in the process of this and behaves accordingly, in ways that make no sense to others. A person can suffer as much from psychosis as they could from a physical disease such as cancer.

These diseases involve direct induction of suffering. By this I mean the source of the pain is within the brain. This can produce a greater amount of suffering than many events that are external to a person. Our perceptions, mood and thoughts are altered by our disease to create a horrible illusory world. The illusions experienced by a person going through psychosis (due to either schizophrenia or bipolar illness) can be horrible and terrifying. These illusions can create a "fight or flight" state of being that never seems to stop. The state of being confused and disconnected from reality that psychotic people get is very hard to go through—it includes an inability to adapt to the actual environment due to the psychotic thoughts that are in charge. On the other hand, the depression experienced by a bipolar or depressed person can be unbearable—to the extent that people sometimes take their own life.

Medication for many of us is a secondary source of suffering due to the side effects. (I will go into more detail about side effects later in this text. However, they are not pleasant.) Getting medicated can be a source of relief for a psychotic person because, to twist a metaphor, one has been taken from "the fire" and has been put back into "the frying pan." Things by no means feel good on medication, but a mentally ill person is usually better off than they were when the disease ran rampant in their system, and created inaccurate thoughts, extreme moods and improper actions.

Thirdly, as people with mental illness, we must deal with the disdain and rejection of those in mainstream society. And this blame and disdain is for something that is actually a medical condition that we have and that we didn't cause. We must often deal with unemployment, since most of those with a major mental illness find work to be more difficult. We must often deal with the departure of friends that we had before we became ill, since many people, when they become mentally ill, lose their friends. We must deal with additional health problems, since the medication we take causes weight gain, diabetes, and other life-threatening conditions. We must often deal with humiliation, since the life of a person with mental illness often contains that. Many persons with mental illness must live in poverty and with a lack of hope.

When someone suffers from cancer, heart disease, or many other physical illnesses, they are said to be brave individuals who "battle" their illnesses, and who sometimes "die bravely" from their diseases. People with a physical illness get a lot of sympathy, support, and caring. They are sent flowers and get well cards. Meanwhile, with a mental illness, people's attitude is: "What's wrong with you? Why don't you snap out of it?" The person is believed to have a character flaw that caused their illness, and they are often treated like dirt for having these diseases. The only difference is that mental illness is a malfunction of the brain rather than of some other organ of the body. Mental illnesses can also be fatal, since they can cause their hosts to die of suicide or an accident precipitated by the disorientation of psychosis. People with mental illnesses do not get a fair shake. When someone has a mental health diagnosis, they are often viewed as shameful by family and society. This is even though the person did not create the illness and they are battling against it, the same as someone would against cancer.

People with mental illness, when they attempt to blend in to society, such as by getting a job, will usually be "in the closet" concerning their condition. It is a label that today gets a person frowned upon much more so than, for example, being LGBT.

Despite life being hard, there are many reasons to have hope.

It is important to remember that there are many good things you will experience in life. While the life of a person with mental illness may be difficult, there will inevitably be good things that will also happen to you. Don't give up. If you follow the guidelines in this book (or if you manage your illness well without using this book), you have a good chance of accomplishing great things in your life. God, the Universe, or Random Chance has dealt us this difficult hand of cards, and it is up to us to know how to play those cards. This book will give   you   some   ideas.

CHAPTER ONE: YOU HAVE A SERIOUS ILLNESS THAT REQUIRES TREATMENT

Mental illnesses such as Schizophrenia, Bipolar, and Clinical Depression are medical conditions that affect the human brain. The causes of these illnesses are not completely understood. However, they are thought to be caused by an imbalance of neurotransmitters in the brain. Schizophrenia is thought to be triggered by excessive dopamine and serotonin in the synapses of certain areas in the brain. This causes part of the brain to have a malfunction. This brain malfunction results "delusions" and in a person becoming split off from reality.

Being schizophrenic doesn't make you less of a person.

Schizophrenia is not a defect of character. It is not a weakness, and it doesn't indicate cowardice. Being schizophrenic doesn't mean your stupid—there have been numerous geniuses and gifted people afflicted with this.

Being schizophrenic is a medical condition, one that you didn't create. You did not give this disease to yourself, and your parents didn't give it to you through any form of mistreatment. You inherited a genetic predisposition for this mental illness, and this, combined with environment (such as during your time in the womb, or perhaps the environment in the public school system) has contributed to your illness.

Schizophrenia is a serious condition.

Schizophrenia is a debilitating and devastating disease that worsens in the absence of treatment. Treatment for this disease is usually medication plus counseling. You should not underestimate the role of either medication or counseling. Both of these together have a chance of helping you get well, of minimizing symptoms, and of making you feel better. The absence of either one of these could prevent your progress and prolong your suffering. Lack of treatment over a long period of time can cause damage to your brain's ability to process information.

A psychotic episode causes lasting damage.

An episode of psychosis, such as occurs to a person with schizophrenia who has refused treatment, does damage to the mind and body, to the brain and soul, and to the life circumstances of the ill person.

What I have said so far doesn't address the damage done to other people. When a person with schizophrenia takes the selfish action of discontinuing medication against medical advice, the actions that get created by the delusions that follow have a good chance of hurting someone else. This includes family members. Family must also stand by and watch as you go through the episode of pain, alienation, and chaos. Family may suffer every bit as much as you do; only they do this on your behalf.

You didn't create your mental illness. No one is to blame for your condition.

Mental illnesses are brain disorders, the causes of which are not fully understood. It is believed to be caused by a combination of genetics and unknown environmental factors, including possibly blows to the head during childhood, and possibly a virus that affects the developing fetus. People have been raised in all sorts of social atmospheres, still with the result that some people become ill and some do not. Difficulties with acquiring social skills during adolescence could be a predictor of mental illnesses that have begun to develop.

Mental illness is not caused by bad parenting.

It is not correct to blame someone, anyone, for your mental illness. Blaming someone will only interfere with your recovery. Unless there are specific incidents of abuse in your past, you should let parents and others in your life off the hook.

There are numerous people who grew up in an abusive, hostile, or difficult environment, and did not become mentally ill. Most persons with mental illness can probably remember some incident of abuse; but then again, many things can happen in life, and most people without a mental illness can probably say the same thing.

There is no cure for mental illness, although there are treatments.

In the 1950's and before, doctors struggled to find some basic understanding of the causes of mental illness. A drug called Thorazine was discovered, and this led doctors to the theory that mental illness is caused by a chemical imbalance in the brain.

Mental Illnesses are Brain Disorders.

Mental illnesses happen to people because the human brain, just like any other organ in the body, is subject to disease in which something goes wrong with this organ. Alzheimer's disease and brain cancer are two other types of diseases that can afflict the human brain. However, the brain is such a sophisticated, multifaceted organ, (the organ of consciousness) that more can go wrong. Schizophrenia, Bipolar and Depression are examples. Diseases of the brain impact human consciousness just as diseases of the heart impact blood flow. The human brain is there for the purpose of guiding the human organism toward adapting to the environment. If something goes wrong with the brain, we will become maladapted to our environment.

A PC computer runs on both its hardware and software. The human brain also runs on its physical structures and on the content of the mind. I like to think that even though my brain has "hardware" problems, it doesn't reflect "the real me." My essence, which may or may not be merely the programming I operate from, compensates for the hardware problem in my brain, and makes me competitive in the world of people. Treatment of my disease through medication allows this to happen.

New Age and Holistic Medicine Won't Solve the Problem

You can't cure this illness with training, meditation, diet or physical exercise. You would merely become a healthier person with mental illness. I am not saying these things aren't worth doing. Yet, schizophrenia is a medical condition that responds to the appropriate medical treatment. You can't cure a thyroid deficiency with diet or exercise, either. Holistic health is just fine but it doesn't address the causes of psychiatric illnesses. You might think, "If I just exercise enough, and eat enough green vegetables, I ought not be mentally ill..." What you get, if you try to treat schizophrenia with exercise and green vegetables, is you'll have a great six pack of abs, and very likely good cholesterol numbers, but without also taking medication, you will be a physically fit psychotic person.

Zen Meditation and Yoga won't eliminate the need for medication and other treatment.

Meditation and other practices of mindfulness work very well to make life more livable, and I highly recommend them. It won't cure mental illness but it will enable you to live a better life alongside having the illness.

When I was in my twenties, I hoped to become spiritually enlightened, believing it might cure my mental illness. What I discovered, after enough meditation and cognitive techniques, was that I could love myself and appreciate my life without having the emotional need to get rid of the mental illness. To me, this could be a more valuable gift.

When Buddhism is applied to the situations included in mental illness, it makes a person be able to adapt to that life situation, and to work positively toward a wellness that is actually more meaningful than being off medication would have been. You discover that life is okay while mentally ill.

CHAPTER TWO: THERE IS HOPE

It is not necessary to settle for a life in which there is nothing to look forward to other than your next cigarette. If you manage the illness and become more well, you have a chance of doing many of those things that you see others do that make life more enjoyable. This includes things like relationships, work, and play.

If you work to make a life for yourself, such as by getting a college education, getting job training, or by doing whatever level of work you can handle at the moment, even if it is just volunteer work, you will experience the rewards of these activities. It is important to remember that if you feel limited, it is only for now.

If your episode of acute mental illness is recent, you should realize that your current level of functioning does not reflect what you will be able to do when you have more recovery under your belt. It may be several, even ten years (in the absence of relapses) before you reach the capacity of functioning that you will have as a recovered person.

Having a schizophrenic illness does not mean that you can not pursue the things you want in life. It doesn't make your life hopeless. It changes things and makes them more challenging, and there will be some things you can not do. However, there will be a lot of things that you can do to make your life better and worthwhile. You may need to retrain for different work than you have done so far, if your previous job does not permit taking medication. An example of this is working as an air traffic controller or airplane pilot. Another two examples (of what you probably can't do while medicated) are military service or joining the police forces. Other positions are more within reach. For example, there is no rule that you could not do some type of administrative, counseling, engineering or data entry work. Most jobs can potentially be done by a person on medication.

I won't lie to you. Being mentally ill and medicated are, to an extent, barriers to working. It makes work harder than it would otherwise be, but it doesn't make all jobs impossible. The more training and expertise you acquire in some field, the more likely it is that you will be able to experience a good employment situation.

Professional employment is a better situation than unskilled work, and is more doable because medication inhibits the ability to move fast, something that unskilled work often requires but that professional employment usually doesn't.

In fact, you can create hopefulness in the absence of evidence—out of thin air. One requirement for this is that you value and believe in yourself. Another factor is pure and simple optimism. If you operate from the assumption that your life will somehow be workable, such an assumption will likely prove itself true.

Having Schizophrenia doesn't mean you're stupid.

Many persons with mental illness are more talented than average people who don't have a psychiatric condition. Schizophrenia sometimes runs in families alongside heredity of higher-than-average intelligence. Albert Einstein had a schizophrenic son. So did science fiction author Kurt Vonnegut. Many persons with schizophrenia are blessed with more talent, but are given the extra challenge of a mental illness.

Persons who are employed in various human services, people who work as caregivers for people with mental illness, may behave toward you as though you lack intelligence. This low esteem from others can be upsetting, especially when we ourselves know that we have intelligence. However, we can learn to look past this error and to instead assume that those who underestimate us are simply ignorant people. Never let mental health staff convince you that you are a stupid person. Once you are convinced of this, you have lost the battle.

Having Schizophrenia doesn't make you a "defective" person.

It might seem to you as though having schizophrenia means that you are some kind of a "defective" or "flawed" person. It is not necessary to view yourself this way. Such a belief concerning oneself has no more accuracy than the erroneous tenets of the NAZI philosophy. It was an ignorant view of people that valued a warped and false idea of perfection.

An automobile that doesn't run properly could accurately be perceived as defective. However, as humans in modern society, we all have problems for which we use some kind of aid. The ones who do well are those who are not afraid to take full advantage of the things that help them succeed. Another way of saying this is that we must all do the best we can with the assets and the difficulties that we are given.

Having Schizophrenia doesn't make you "weak," "sick," or "of bad character."

A mental illness such as schizophrenia carries no intrinsic moral content, good or bad. Schizophrenia is a medical condition unrelated to a person's character or merit. It is a medical condition, pure and simple. It carries no value of right or wrong any more than does an ingrown toenail, or perhaps a thyroid deficiency. It is not an indication that you are weak. Having a mental illness is unrelated to your level of worthiness as a person.

If you were a sociopath, I would be saying something different. For example, chronic shoplifting, and doing this when you are aware of what you're doing, might show that there is an important ethical piece that is missing. Premeditated murder is another example of a person who puts their self narcissistically above the need to be kind to others.

Having Schizophrenia doesn't mean that your life is over.

Having schizophrenia doesn't mean that you can't do something useful with your life.

Before I chose writing as my life work, I did electronic repair and was sometimes pretty good at it. When I felt that many of the repair jobs were too demanding for me, I went into business for myself. It was my belief that I could tailor the structure of the business to suit me with the strengths and weaknesses as well as preferences that I have.

Self employment should be seriously considered by you if you are intelligent and have a mental

illness.   It is a way of departing from the
productivity or speed requirements that companies
may have.   At the same time, if you are good at
what you do, you could potentially do very well as a
self-employed person.   It isn't necessary to hire
anyone.   The red tape involved in a sole
proprietorship is minimal, and there are numerous
books out concerning home-based businesses.

CHAPTER THREE: BEGINNING YOUR RECOVERY

COMPLIANCE AND NONCOMPLIANCE:

Messing with your medication without doctor's approval, or going on and then off medication is like playing Russian-Roulette with your brain.

 It is only a fool's game to try to be your own doctor, mess with the dosage of medication, go off medication on your own, or go off and on medication. The brain doesn't react well to being on an incorrect dosage of medication, or to sudden changes in dosage. A psychiatrist needs to be consulted before changing anything. It is a misconception to think you can "tough out" a psychotic episode and "get to the other side." If you try that, you will ultimately discover that this "other side" of a psychotic episode doesn't exist. The person with mental illness who believes they can cure their self ends up getting extremely sick, and experiences suffering on an extreme level that is horrible to go through. You will just keep getting sicker until you are intervened upon with medicine. That is the usual outcome. I have never heard of someone successfully going off medication. Ideas that go along with Holistic Health, such as curing yourself with nutrition, or with the power of mind over matter, are ideas that sometimes kill people, such as cancer patients who would otherwise accept conventional medical care. In general, can't cure mental illness with nutrition. Unless you are the one in about a half million people diagnosed with schizophrenia who have been misdiagnosed, and who are psychotic due to a severe food allergy, you should stick to conventional treatment for your condition.

Don't let your delusions fool you into believing that you can get by without medication. "Noncompliance" with taking medication to treat your mental illness will usually end up with you getting sick all over again and returning to the inpatient psych unit.

Too many repeat episodes due to noncompliance can cause a form of brain damage.

When among other persons with a mental health diagnosis, have you noticed that people seem dull-witted? This may not be due to the medication they take. I take antipsychotic medication in a very substantial dosage, and no one has accused me of being dull-witted. No—the reason people seem cognitively foggy is probably that these people have experienced numerous repeat episodes of psychosis, and this has given organic brain damage to these people. Following the episodes of psychosis my recovery time has increased to more years with each successive episode. In other words, after my first episode I recovered in less than a year. Episode number two took a couple of years before I felt back to normal. Episode number three actually took five years to recover from. And my last episode of psychosis (and I hope to God never to have another one) took me ten years before I felt at my normal level of functioning. Although there are some aspects of my functioning that may never come back.

Imagine someone who has three psychotic episodes in three years. The amount of shock that the brain of such a person must go through can easily cause that person to become "dull-witted," as many persons with mental illness seem to be.

You should consult with a doctor prior to changing your dosages or your medications. You should get doctor's approval before stopping or changing dosages of a medication.

If a medication you're taking is new, and you discover that it has unbearable side effects or that it might be making you ill, you should really check with your doctor before discontinuing it.

Above all, don't keep secrets. Nothing is worse than being noncompliant and not being honest about it. If you tell your doctor that you discontinued something because it was unbearable, it gives him a chance to prescribe something else for you. If you aren't forthright about it, you have a good chance of falling through the cracks.

Mental illness is for real and wasn't created by the drug companies to help them sell pills.

Some skeptical people try to say that mental illness isn't prevalent in third world countries that don't have medication available. It is an argument in favor of not taking medication in the erroneous belief that the brain will correct itself.

First of all, it is not accurate that mental illnesses don't exist in third world countries. The misleading appearance of this is caused partly by a lack of ability to collect good statistics in those countries. People who become mentally ill in a place where there is no doctor, probably die due to being unable to meet their basic needs. This could certainly change the statistics by reducing the number of mentally ill in their censuses. They died and weren't counted. My wife visited Kenya when she was in college, and told me that mentally ill people would just wander around aimlessly and could not survive for very long.

Before medications were widely used in the US, persons with mental illnesses apparently were forced to live in cruel asylums, and there was very little that doctors could do to help them. When Thorazine was discovered, it was like a miracle drug on a par with Penicillin. Prior to this, doctors would perform lobotomies, electroshock, and insulin shock on people. Medication, including instances when it is administered by force, seems to be a far more humane alternative compared to the way persons with mental illnesses were once treated.

Mental illnesses probably take place when the "self correcting" nature of the brain has become impaired. It is probably true that most people's brains are inherently self correcting. However, people probably become ill to begin with because something has gone wrong with their system of self correction. Thus, that person needs chemical intervention.

Every organ in the human body has diseases that are associated with it. The human brain is no different. It is subject to Alzheimer's, stroke, epilepsy, concussion, and other diseases, including mental illnesses. The human brain is the most complex organ in the body; therefore it makes sense that more things can go wrong with it.

With treatment, over time, the brain of a mentally ill person can get healthier. This is reflected in such a person's experience of life. People have a chance to feel better.

If you have a full-blown psychotic episode and are then given treatment that brings you back from it, you will probably not feel like your usual self for another several months following recovery from the episode. A psychotic episode brings a form of brain trauma. It is not the same as either a physical blow to the head, or deprivation of oxygen to the brain. However, the damage is real.

In order for you to recover your mental faculties, you may need to exercise your brain through activities in which you use your mind. This is not limited to reading and/or studying. Action, physical activity, uses other mental capacities that aren't employed very much when performing academic work. It takes brain power of a different sort to make an omelet, to wash dishes, or to do a load of laundry. If you're looking for "manlier" work, you could go out and mow a lawn. Physical activities like these exercise the motor skills in the brain, and get you used to performing actions. Sitting on the sofa and watching television doesn't employ the powers of the mind very much and doesn't qualify as exercise for the mind. And of course, you ought not to leave out reading books and articles, or reading posts on the internet. Messing with computers, turning nonworking computers into working computers, is another good way of exercising the mind.

You will find that if you don't exercise these capacities, your overall mental condition may fail to improve by very much. When you injure a leg, part of the rehab of that leg includes exercise when you reach a point where it is partway healed. A psychotic episode is like an injury to the brain. It is a trauma to the brain because the change in chemistry is a shock to the brain cells. This means that functioning is lost, and work must often be done to regain that functioning. The brain responds to exercise in the same way as the muscles in the rest of your body. With exercise, the brain becomes stronger and better. This requires effort.

In the first one to two years of recovery that follow a psychotic episode, you may not feel in your best condition. As time passes, and with a routine of activities that exercise the brain, you should begin to feel closer to normal by year three.

A sign (among others) of the brain doing well is that, ironically, you will encounter problems. A normal person has problems in life that he or she must deal with on a regular basis. There is no "happily ever after" that you think you could get as soon as you're more recovered, or, as soon as you have achieved whatever goal you are thinking about. Life continues to be difficult, even when the illness is well under control. And the fact that you may find life to be hard will not be abnormal.

The ability to enjoy life is like a prize possession that ought to be protected. I have come to disagree with the new age philosophies, the ones that are cult like spinoffs on Buddhism that state you ought to be happy all of the time. If you are getting well, you will have bad and good moments.

DEALING WITH SYMPTOMS AND WORKING WITH YOUR PSYCHIATRIST

A psychiatrist is a medical doctor who can help you receive appropriate treatment for schizophrenic or other symptoms. Not all psychiatrists are the same.

There are some psychiatrists with whom I would never be able to get along. When a psychiatrist is found that's good for you, you should work with him or her to arrive at the best possible treatments for your condition. You should give him or her full and complete reports of your symptoms to the best of your ability, and you should not keep secrets from him or her. A good psychiatrist will treat you with some amount of respect and dignity. Such a doctor can work with you to arrive at the right medications. A good psychiatrist cares about you and doesn't want you to suffer either from unbearable side-effects, or out-of-control symptoms.

A good psychiatrist can be like a mentor, and you may form a bond with him or her.

Your psychiatrist is an imperfect person. Mine once advised me to give up on my illusions of what writing would do for me. I disregarded that piece of advice and went on to become more successful at writing than I was before. You do not always need to let treatment professionals shoot down your goals, especially if you really believe that you have ability.

Treatment professionals may try to shoot down your goals if they believe that you are having delusions of grandeur. There is a fine line between giving up unrealistic plans, versus being determined to achieve something that you believe you can do.

In terms of knowledge and expertise in the medical arena and pertaining to your medication, you should let your psychiatrist be the expert. (This is assuming you didn't get a bad apple. If you feel badly treated by a psychiatrist, it could be a signal that you need to change doctors. It is important to see treatment professionals with whom you feel compatible.)

You are the expert on how you feel on a given treatment regimen. No one else knows you from the inside out. They can only see your external self—no one else knows your thoughts. No one can read your mind. It is up to you to report your thoughts and your feelings to your doctor, so that he or she can best help you.

Learn to recognize and counteract delusions with your conscious mind, as an adjunct to medication.

If you are experiencing a thought and are wondering whether or not it is a delusion, versus a "normal" thought, you can share that thought with someone you trust, who doesn't have a mental illness, and ask that person if they believe your thought is a delusion. This is one type of "reality checking" wherein you "check reality" to measure the accuracy of thoughts.

If you have a delusion, there is no shame in that. It is simply your brain's malfunction and it is also meaningless. There is no point in trying to discern if there is some hidden message in your delusions. Such a thought process, in which you are trying to find hidden meaning within the nonsense of delusions, would be detrimental to your mental health.

Your mind is like a sophisticated test instrument which is designed to create a picture of your environment. Any test instrument, whether used by a technician, a scientist, or a novice, must be properly calibrated if it is to be accurate. And, just as a test instrument must be accurately calibrated to be useful, your consciousness must be "tuned" to create a usable picture of your environment. Most people do this through social avenues. People usually look to their fellow human being to get information that is used to adjust their consciousness. This is also called, "living in 'common consent' reality." If your mind has the same picture or a similar picture of the environment as do others, you will not be seen as an "insane" person.

When people exclude other people's viewpoints, it entails more of a risk. Sure, some of our society's most brilliant and most accomplished people thought independently and disbelieved what others believed. However, most of those who depart from commonly accepted reality find themselves to be clueless, and off on the wrong track. The important thing is in knowing when and where you can disagree, and to do your disagreeing in a socially acceptable manner.

*   *   *

A person with low level psychosis can learn to use cognitive techniques to clean up some of the thoughts. To do this, one questionable thought is evaluated at any given time. You can evaluate a thought and ask yourself if you think it is something your psychiatrist would believe, or would instead label as a delusion. You can evaluate the thought based upon your individual knowledge of what types of delusions you have had in the past. If, in the past, you had delusions about extraterrestrials and flying saucers, then more thoughts about them ought to raise a red flag. If a thought makes you special in some way, it is more likely to be a delusion than a thought that makes you essentially the same as other people.

With effort, medication, compliance and persistence you will probably feel better and better, the longer you remain stabilized and out of the hospital.

I suggest a proactive effort at recovery. A proactive recovery means that you will use effort to do things that make you better. This includes mental and physical activity, remaining on a consistent medication regimen, and putting effort and consciousness into your activities. Using effort at things rather than doing the minimum allows you to get mentally stronger.

CHAPTER FOUR: REMAINING STABILIZED

After a psychotic episode, perhaps a few years after, the memory of how badly you suffered during the episode has faded. Also, the memory may have faded of the extreme level of power of the symptoms of this illness. Not remembering how bad it was can make you tempted to try noncompliance once again, and this is foolish. How many times do you need to be bit by the same dog?

Repeat episodes should be avoided.

In a repeat psychotic episode which is often triggered by stopping medication against medical advice, one of the things that take place is the withdrawing of medication and its associated rebound. This adds a lot to the force you are working against when trying to do without medication. In addition, you are working against the original problem that made you get put on medication.

You are not a "sick person" because of this. While your brain may have a problem, it doesn't mean that "the real you" is sick—it is forced to take a back seat in your functioning because you have been involuntarily taken over by the delusions.

I must emphasize here: Going off medication against medical advice is foolish and dangerous. Unless your psychiatrist's name is Frankenstein, he or she probably has some amount of expertise that should inspire you to listen. This is not to say that there are no bad psychiatrists. If you think you have a dud, rather than simply disregarding their instructions, you should find another doctor and get a second opinion. Stopping medication can lead to a repeat episode of psychosis. You do not want this to happen.

When I was eighteen, I went off medication against medical advice, and the illness caught up with me a year later. In subsequent attempts to do without medication, it took less time for me to become ill and it took me longer to recover once medication was reintroduced. This could have been due to the illness worsening as I got older, as well as the rebound effect of stopping medication. Had I remained compliant, I would probably now be in a much better position in life. A great deal of your time, effort and work will go "down the drain" in a repeat psychotic episode.

Your time should usually be structured.

You are better off if you have some activity out of the house to go to every day. In my case, my activity of writing is in the home. However, I treat it as a job, one in which I am "on the job" for several hours each day. I have even begun a ritual of having "work lunches" such as a can of chili, or a bowl of ramen with vegetables. I also go on daily trips either to the post office, or sometimes Starbuck's.

Without an activity to look forward to each day, your condition could worsen. If you have chosen an online university as your activity, it is important to get a support network that encourages your efforts. As a writer, I am happy when friends and family ask how the writing is going, and this validates what I am doing.

Other than the possibility of attending an online university or online class, your activity will probably be out of the house. The structure that you choose should involve meaningful contacts with people. In an online class, you are dealing with your instructor and with fellow students. In a volunteer job, you will see other people in person with whom you will interact. The interaction is important. Interacting with other human beings, especially those who do not suffer from a mental illness, helps keep your mind on the right track.

If the "meaningful activity" that you choose is unrelated to the mental health system, it will raise your self-esteem more than an activity intended for persons with mental illness. There is no shame in having a mental illness. However, if you can do something every day that brings you into contact with "normal" people, it will help you. Even if your contacts are mostly online, that still counts as interaction with people. Hopefully the activity you choose is constructive and helps people. Volunteering for a local animal shelter qualifies. Playing online poker games, or else viewing pornography on the web, are two things that don't count as valid daily activities. The activity you choose should stimulate the intellect, it should get you around people, and it should be positive.

If you are stuck waiting in a waiting room because you are about to see a mental health practitioner, or someone else, check to see if there are magazines, and if there are, pick one up and look for an article that interests you. Reading while waiting will stimulate your mind, give you knowledge you would not otherwise have, and it will make a better impression. The alternative is to sit like a zombie, or maybe, play with your phone, either of which comes across as mindless.

You should take your medication as prescribed.

In this book, I emphasize, perhaps redundantly, that it is important to take medication, and to take it according to doctor's instructions. This is the only way that I know of to support a real and meaningful recovery and to prevent devastating relapses. Opponents of medication are not thinking in real terms. Their belief systems are barbaric and superstitious. When you defy advances in physical science, ones that are closely connected to realities that anyone can observe, then you have invented your own religion. Medication makes you better. Some medications are bad for you, and some are good. To try to do without medication, with these illnesses, it is the equivalent of thinking you can walk on water.

The issue is made more difficult by the abuses that have taken place in mental health treatment. Cruel treatment of persons with mental illnesses, such as putting people in four point restraints, giving people inappropriately high dosages of medication, and depriving people of basic dignity are all things that have been perpetrated by the ones who also promote medications. It is only natural that we would want to defy any axiom that comes from those perpetrators.

It is fine to resent abusive people. The world is a mixture of truths, lies, and half-truths, and these all come from the good, the bad, and the ugly. Don't let your justified resentment deprive you of medicine that could do you good.

The best revenge that you can get upon the perpetrators of cruel actions in the mental health treatment system is to take your medication and get well enough that you will never again be in a position of being under the control of those evil people. It is at that point that you can further your revenge by speaking out. These are legitimate avenues of evening the score that are accepted by society and that don't have negative repercussions upon you. Our constitution continues to guarantee freedom of speech and expression with a few exceptions. You are not allowed to yell, "Fire!" in a crowded theater, unless there really is a fire. You are not allowed to inaccurately publish bad things about a private citizen, as this is called "libel." You are allowed to speak of what happened to you, although it is advisable not to use people's names other than your own when you do so.

If you don't take your medication and get well, you may never be in a position to do any of the above things. Violent or illegal acts of revenge will only perpetuate the system of inequality that continues to make persons with mental illness suffer.

Exercising your mental capacities will help you in life.

When you have become stabilized on medication, and when things in life are somewhat smoothed out, you should perform mental exercise to increase your mental strength and capacity. Mental exercise, as I am calling it, doesn't exclude physical movement, as in performing tasks. Some of your mental exercise should involve regular reading, and this can be done while on the internet, or the old-fashioned way by reading a physical book, newspaper or magazine.

If you do not exercise your brain cells, they will shrink and become useless. This is especially important for people who take antipsychotic medication. Antipsychotic medication makes you not feel like reading, and may make you tempted to sit like a zombie. However, following such a temptation, and failing to overcome the difficulty reading, or being inactive in general, will cause atrophy of the body and mind. You will lose your ability to function.

Reading, writing and thinking are the mental equivalent of a bodybuilder's workout. If your mind is weak, everyone will take advantage of you and you will not succeed in life. Your brain will become a strong ally, and you will have a shot at getting what you want in life, if you make your mind strong and capable through mental workouts. This includes physical tasks, such as cooking and cleaning, yoga, or physical tasks in a job; physical tasks with some complexity exercise the brain as well.

CHAPTER FIVE: CONSEQUENCES OF BEHAVIOR

Denial, if you have it, is something you believe you don't have.

The Buddhist Law of Karma is a similar idea to Newtonian Physics where every action has an equal and opposite reaction. Karma says that the force we put out into the universe comes back to us. In no way does Karma say that we are being punished.

Karma is like you are entering data or a command into an infinitely powerful computer system, and the response is usually "over our head" but is related to the initial query that we entered.

What does this have to do with being schizophrenic?--you might ask. The actions that we take when we are delusional, or even those we take when we are merely in denial, will come back to us. Being delusional is one thing; a person is partly but not completely responsible. Denial is something else. It is common for someone with schizophrenia to experience a lot of denial when medicated and stabilized because this is a lower level of being delusional. The universe doesn't care that you were messed up when you did the thing you did. It simply sends you back the energy you gave it, and it does so with no moral judgment.

The Buddhist path of some people has a stage of "paying back your Karma." Paying back your Karma for a recovered schizophrenic takes place upon finding clarity that replaces the delusions. Once clarity is found, it becomes possible for us to look back on our history and see how messed-up, without at the time realizing it, we were. It is at that point that we are ready to apologize to the world.

This usually consists of a series of uncomfortable but probably harmless events in our lives. Not that we need to do this to ourselves in a deliberate manner—doing that would produce more bad Karma. Leave it up to the universe to furnish you with the difficult situations that you must resolve. It this never happens, then, lucky you!

A person in denial believes that the rules of the world that govern everyone don't apply to him or her. If a person behaves a certain way, let's say offensively, people will be offended whether they tell you this to your face, or not. More than likely, behaving offensively results in people talking to one another about you and about how gauche you are, and they will do this without bothering to tell you about it. This is an unfortunate trait of human behavior; people are the last to know about something they have done wrong. It is up to you to attempt to figure out how others were affected by something you did.

The problem we have as people with schizophrenia is that to get well, we are taught to take people by their word. This is an important mechanism in our recovery, and it prevents us from becoming delusional even though we are also taking medication. However, people who don't have schizophrenia are taught to believe they are being lied to, or that they're not being told a foremost point that they are supposed to be getting. Thus, recovered schizophrenics are naïve by necessity.

For something you did while delusional, you may be negatively judged for years or decades afterward. This isn't fair, but it is so.

Since there may be people who don't approve of you, it is important that you approve of yourself. There was a man who recorded an audiotape on success and the fear of success who said something to the effect of, "You are a successful person as soon as you grant yourself self-approval." In life, not everyone is going to like you or approve of you. It is important to buffer yourself against those who are negative about you—buffer yourself with positive thoughts concerning yourself. This creates fortitude. You are fortifying yourself with self-promoting thoughts. If you are not willing to take your own side, who will?

Bad behavior often has bad consequences; good behavior often has good consequences.

If you produce some type of work on a regular basis, you will probably at some point reap some kind of reward for this. If you want something, it doesn't automatically follow that you deserve it. Wanting something, no matter how much, does not equal earning something.

Furthermore, being financially rewarded is not always the outcome of performing hard work. No one can predict the future. You could start a business, nurture its progress and work hard for many years in the hope of being successful. The outcome of all that work is sometimes being exhausted and penniless.

Performing the work involved is normally a prerequisite for success but is never a guarantee of it.

If there is something you want in life, such as money, a big home, cars, or even relationships, you should realize that you may not get the thing you want. Your desire for something exists only within your mind. You should balance a desire with the thought that you will be okay whether you get the thing you want, or not.

*   *   *

Getting violent, and / or going off medication against medical advice are two behaviors that should never be an option. You probably can't predict how lack of medication will impact you, or how you will behave as a result. If off medication your delusions could make you do something illegal, and someone, possibly you, could be harmed. Going off medication is like Russian-Roulette; you might survive and you might not.

That said, the author is guilty of going off medication against medical advice in the past. The last time I tried that was in 1996, and the result of it was becoming psychotic, with the accompanying "wild" behavior, and then re-hospitalization, followed by starting over again at square one.

Going off medication, unless a doctor orders it, will often result in getting completely psychotic. Upon recovery from this, a great deal of progress in many areas will be lost.

Some of the negative effects of past psychotic episodes are cumulative. With too many psychotic episodes in one's past, one loses areas of functioning. Have you seen mentally ill people who seem to be dull-witted? This could be the result of having one too many psychotic episodes with the resulting brain damage.

CHAPTER SIX: HAVING A GOAL

Acknowledging a mental illness can seem like a disheartening thing to do. After all, beforehand you may have had great ambitions in life, and may have wanted to do many things that may now seem out of reach. Without mental illness you may have felt optimistic about life. Now, people in your life could be telling you that you can't work.

When I was twenty years old, a counselor wrote comments in a progress log that said: "Jack has not yet acknowledged that he can't work." Following that, in my twenties, I worked quite a bit, and this included being a technician and manager at a television repair shop. I worked at quite a number of jobs in my twenties and I succeeded at some of those jobs. And this proved wrong the counselor who assumed that I would not be able to work.

You should not always let people tell you that you can't. In some instances, you really can't do something. A job in law enforcement will probably not work if you are on medication. The same goes for trying to become an air traffic controller. Certain jobs simply can not be performed by a person who must take medication. However, most jobs can be performed by someone on psychiatric drugs. You won't know until you just try.

Having come close to fifty and having tried to work at a number of jobs, I feel burned out on employment. In my case, because of the illness, the medication, and the wear and tear caused by both, I have opted to stop attempting regular employment. However, for me, that does not rule out starting my own business. At some future point, I intend to start some sort of technical assistance company.

Choosing a goal that is realistically achievable for you is important. To achieve something, break it down into its component parts.

Here are some examples. To start a technical assistance company, you may first want to gain some knowledge of the things you're going to work with. If you intend to specialize in work on PC's, then there are two kinds of certificates that you might want to consider. One of them is called the "A+ Certification" which involves passing a written test. Similarly there is something called, "MCSE" which is a bit more advanced. Electronics training of some kind is necessary. And if you expect to become self-employed, business training will be helpful. Before becoming self employed at something, it is useful to work for a successful company in that field in order to learn how people are doing things. If you do not learn the industry standard practices in a field of work, you will be working at a vast knowledge deficit compared to your competition. Therefore, elements of starting a business include training in your field as well as learning what your competitors are doing. If there is an area of work that you have done in the past, that you are good at, then you might want to consider becoming self employed at that.

Even if you work for a company for a short time because that is all you can manage, you will have learned much from that experience.

For the remainder of this chapter, I thought it would be useful to include an edition of my column from The Berkeley Daily Planet that I think applies to this subject:

ON MENTAL ILLNESS: PROSPECTS VS. LIMITATIONS

When a person with mental illness, their family, or their treatment practitioners anticipate few prospects toward a career or other goal, it may be unnecessarily pessimistic. It can also become a self-fulfilling prophecy. Persons with mental illness and their families should have hope tempered with realism of a possible good outcome. Having optimism and the preparedness to try increases a person's chances that they will do something good in life.

For those who hoped to do work of some kind and make a future for themselves, maybe encouragement would be better than premature hopelessness. For those who merely want some relief from their illness and to whom success doesn't matter as much, they should not be pushed into excessively hard work.

If someone with a mental illness does not feel a pressing need to prove something in the world, and would rather focus on his or her recovery, they should not feel pressure to "do something" coming from this manuscript. The idea of "accomplishing something" (such as a job, a hobby, volunteer work, school) is only a suggestion and comes from an agenda that not everyone needs to follow.

Depending on the severity of someone's illness and the level of his or her recovery, a person with mental illness can end up relatively capable or relatively limited. This sometimes translates into limited prospects in life.

It can be depressing to face a disability that removes you from an arena in which you were once successful, or that prevents you from, in the first place, competing. It is not easy to face having limited prospects.

A person who faces a major mental illness undoubtedly must live with some restrictions. Trying to act "as if" there were no disability could lead to a relapse. For example, adequate sleep for a person with mental illness is a must. Most persons with mental illness seem to also require more idle time than others. The things we must do to maintain wellness may draw on a considerable amount of time and energy.

I had problems maintaining work in my twenties, and I realize in retrospect that it was largely a psychological problem rather than a limitation from the illness. I found myself in a negative pattern that I did not know how to escape. Not grasping the realities of work situations and not having enough clarity were factors.

For one thing, I was choosing the wrong jobs. In my twenties it was relatively easy for me to get hired, but it was hard to do the work. I was unclear about the fact that in order to work at a job, I had to tolerate a significant level of discomfort. My emotional equipment wasn't working well enough to do that.

When trying to gain success in the face of obstacles it is important to have clarity about what one is up against. A person can accomplish so much more when they do mental exercises or meditation to get the mind on the right track. This allows one's efforts to be grounded in reality. Thus you are working with people, places or things not as you imagine them, but instead, as they are.

Numerous persons with a mental illness can do an honest self-assessment to determine where they have talent, and to determine in what areas they are more limited. Success is more likely to come from pushing the envelope in the areas that seem to work better. Had I not chosen writing, I would doubtless be doing or trying to do some kind of electronics or technical job. Most likely I would be self employed at this because the regular jobs in that industry are nearly all full-time, which is beyond my limit.

If a person with a disability does not feel able to work at a nine-to-five job, and feels unprepared for several years of going to college, it does not rule out all accomplishment. A person could start a small business in an area within their forte.

Becoming self-employed may be a better situation than regular employment for many people who follow a different drummer. In self-employment, it's good to pick a type of business that seems worth doing, even without the anticipated money. In today's economy, self employment is less promising than in the past and should perhaps be done with a goal of breaking even, and for the purpose of obtaining other benefits of being in business.

The good thing about self employment for persons with disabilities is that you can tailor the structure of the business to suit your strengths and weaknesses.

As an example of small business; starting a computer services business can be a very low overhead situation, since you do not have to rent a space, and you are only paying for advertising plus administrative costs. Unfortunately, a lot of people are doing this job; consequently you may need some kind of additional angle to get customers. Even getting one customer per week could make a significant augment to an SSI budget.

It helps not to take success or failure too seriously. Excessive work ethic is bad for persons with mental illness (and those without) and can create illness or other problems related to stress. A person with mental illness should not be pushed to their extremes in the hope of excellence. Many persons with mental illness, if pushed excessively in the name of work ethic, end up having a meltdown rather than delivering more work.

When dealing with a person with mental illness who has goals it's important not to make remarks that detract. It is hard enough for a disabled person to try to achieve their goal without needing to be discouraged on top of it all.

CHAPTER SEVEN: SELF ESTEEM AND REASONS TO LIKE YOURSELF

Having a mental illness can lower a person's self esteem because they may incorrectly believe they are a "defective" person, or that they are somehow less of a person than someone who doesn't have an illness and who works nine-to-five. It is very difficult to be happy if you have low self-esteem.

Your appreciation of yourself, something you deserve, is a gift that you can give to yourself—through mental exercises, and without the need to go prove something. Get a paper notebook and start listing things that are good about you. Release yourself from self-hate concerning what you perceive as your deficiencies.

I thought it would be nice to include a column on this subject that I wrote for the Berkeley Daily Planet. The following article seems to cover this subject fairly well:

ON MENTAL ILLNESS: ACCEPTING ONESELF WITH IMPERFECTIONS

It can be hard for people with mental illness to accept the idea that there is a "defect" in their brain. This is one reason, among others, why some people are in denial of the illness—there is a conflict between liking oneself, versus acknowledging what seems like a significant flaw. Newly diagnosed people must come to terms with the idea that they may have this biological "difference."

It is important to distinguish between a: your attitude, which is something you can control, versus b: something you can't control, namely, the fact of having a psychiatric illness. Your rating in life isn't necessarily based on the cards you've been dealt, such as being dealt a human body with some problems, but is instead based upon how you play those cards (in other words, where you go in life despite this circumstance.) Another way of saying this is that it doesn't matter so much how good your brain is, but how well you use your brain.

I didn't manufacture my brain, my genes or my environment. Therefore, I did not bring my illness on myself; my illness is not my fault. However, I do have a great deal of power over how I behave when medicated and stabilized. The fallout of my actions deservedly comes to me. But I can still accept myself as I am.

For many people, self esteem is related to a distorted version of Darwinism. People use the idea of "survival of the fittest," as a standard against which to compare themselves. Believing that you could survive if you were put on a deserted island, or that you can propagate your seed better than others, or that you do not need to rely on any assistance from modern medicine to survive, is an arrogant version of buffoonery. Prior to modern advances, people rarely lived beyond age 30, and most died a lot sooner than that due to diseases or to other harsh conditions that once existed. Today, most people rely on some kind of help from modern medicine and modern technology.

Accepting the help of technology rather than foolishly trying to do without shows maturity and means that I am not a masochist.

I am not my brain. I am the person, maybe the personality, or perhaps the entity that is using my brain. It is not necessary to believe I have a "spirit" or "an immortal soul" for me to not identify with my brain. I can see myself as an abstraction which is the final output of my brain, my environment and the software that runs my brain. You could call this an agnostic version of spirit. My [agnostic] spirit can recover from the delusions or other errors that my brain produces and can produce reasoning that has common sense.

I am not "a schizophrenic," or "a bipolar," I am a person, a human being, who <u>has</u> schizophrenia or who <u>has</u> bipolar. It is a form of bigotry to identify us by our diagnoses. This includes when we do that to ourselves.

■■■■■■■■■■■■■■■■■■■■■■■■■■■■■■■■■■■■■■■

Again, it is a good idea to get a notebook and make a list of reasons why you are a good person and why you ought to be proud of yourself. You need not compare yourself and your accomplishments to those of non-afflicted family members. It is ok to rate yourself in comparison to other people who also have a mental illness. Mental illness is a major handicap and should be accounted for in any assessment that you do of yourself.

I have a friend who doesn't understand the difficulty introduced by my mental illness, and we argue a lot. That person believes I have limitless capabilities and that I am not trying hard enough. Other friends have more comprehension of the fact that mental illness is a real disability. It is unfair when someone expects us to be as good as "normal." This includes instances when we do that to ourselves. It might seem like a form of bigotry to assume that a person can do less if they have a mental illness—however, I think it is not. We are acknowledging a difficult reality.

Most persons with mental illness are "normal" when you look into our hearts. We are "normal" in the things that we want in life. Our basic consciousness or our sentience is "normal." However, our capabilities are often less than what a "normal" person can do. When you look at someone in a wheelchair, it is unfair to tell them they need to get up and walk. It is the same situation with those with a mental illness, only the wheelchair is invisible.

Assess yourself in comparison to others with mental illness and appreciate yourself for your capabilities and not your limitations.

CHAPTER EIGHT: RELATIONSHIPS

Shyness and social awkwardness are big barriers to initiating a romantic relationship. Someone can be shy and awkward under certain conditions, and can be outgoing and confident under other conditions. Many persons with schizophrenia have difficulty being social. If that is you, the first thing to remember is not to play self-hating scripts in your thinking over it. Do not punish yourself with negative thoughts just because you have some shyness.

People still manage to get relationships even if they have shyness or other idiosyncrasies that get in the way. If someone is friendly to you, and you are interested, I understand it is good to start with small talk. You could bring up the weather or perhaps the price of gasoline.

I admit that I'm not an expert on this. I will say that despite being fairly shy, I have had my share of relationships, and then, about sixteen years ago, got married to the right woman.

However, relationships are more than a romantic partnership. Relationships encompass how we relate to anyone.

Relationships, including those with members of your family of origin, are a give and take. You may sometimes find that you must compromise with someone, even if you believe it is unfair and that you are right.

Relationships have the power to heal.

Since we all live on the same planet, anything we do in some way affects everyone else. It can be said that we have "a relationship" of some kind with everyone on Earth. The variable is how involved we are and what kind of involvement. If someone is chosen at random on the other side of the globe, their climate might be affected by the size of my carbon footprint, and their employment might be affected by any item that I purchase.

Furthermore, since my body has mass, the person on the other side of the planet is affected by changes in their gravitational field any time I walk across the room. We all affect one another. Everyone has some kind of relationship with everyone else.

In some instances, the relationship that you have with someone consists of mutual animosity. Your actions still affect the other person.

It is up to you and other people how involved you will become and what type of involvement. You can still offer a person your respect as a fellow human being, as well as your emotional acceptance, although you might never see such a person again. Or, that person could be your sworn enemy.

In dealing with relationships, learning how to let go of a person is a very good skill to have in your repertoire. If things are not mutually beneficial, being able to let go of a person, on an emotional level, will save you a tremendous headache. If someone is bad for you, or if they have said they don't want you, it is time to learn to love that person at a distance. Or don't love them. Either way, if someone has shown that they don't like your presence, don't punish yourself by hanging on.

It might seem like the end of the world to let go of a person with whom you are enamored; so be it, let the world end. There is a life waiting for you after you let go of someone who is bad for you.

CHAPTER NINE: SIDE EFFECTS OF MEDICATION

Side effects of psychiatric medications are one of the chief causes of "noncompliance" among young people with schizophrenia. Taking medication can be very uncomfortable. Side effects can create a lot of suffering and can make a person feel like crap. Side effects can include stiffness, restlessness, a drugged sensation, dry mouth, and sometimes can include Tardive Dyskinesia.

My first outpatient psychiatrist, who I will call "Doctor M," explained Tardive Dyskinesia to me. It is an involuntary movement in the mouth, face, head and upper body. If medication is discontinued, sometimes it only worsens the side effect. Thus, it is something that can be irreversible including instances where the medication is quitted. It is not a great motivator for taking antipsychotic medication.

Tardive Dyskinesia supposedly occurs in about one in one hundred cases. Thus, for any one person, it is fairly unlikely that you will get it. I have seen people suffering from this side effect, and these individuals are still going on with their lives—it is not the end of a person.

I believe taking antipsychotic medication is important enough to brave the risk of Tardive Dyskinesia. If the medication isn't taken, what sort of chance at life does one have? Not much. It is worth the risk unless a person's psychosis is slight enough that they can get by without meds. In my case, my psychosis has been very severe any time that I tried to do without medication. Thus, I have no real choice concerning taking medication, and for me, Tardive Dyskinesia will always be a risk.

There is no easy answer to the predicament of side effects, although there are some ideas that may help keep suffering at a minimum. Meditation and physical exercise are two things that can help with the stiffness, restlessness and the drugged feeling brought on by medication. Dry mouth can be addressed with hard candy, water, diet soda or chewing gum. You may also get blurred vision, which can be helped with reading glasses. These adjustments might seem outrageous if you are young and are unused to these sorts of things. However, this is better than wandering the streets in a psychotic mode.

The more years that you're stabilized on medication, the more progress you can make in life. Your brain condition should improve, and you will have a chance to make more progress in your life situation, in relationships and in careers. As time goes on, you won't notice side effects as much. As a result, side effects of medication will become less important to you. In short, it is worth it to take medication and feel "like crap" rather than to be psychotic and not have a real chance at life.

CHAPTER TEN: HOUSING

Finding and keeping good housing seems to be a universal source of difficulty for persons with mental illness. In some housing situations, such as those set aside for psychiatric consumers, we are harassed by other tenants who may or may not also have a mental illness. When I was in housing for persons with mental illness, people would come to my door at three in the morning and ask for cigarettes. A man who I would not give a lift to the junkyard for auto parts took it personally and started a fistfight with me.

If living in low income housing that isn't set aside for persons with mental illness, you could be dealing with people who are even more dangerous. In places where it is easy and inexpensive to rent, it is generally not worth renting there. Some of the areas are drug and gang infested. Living in a low income area can be a risk to your life.

Having a good income and a good credit rating would allow you to live in a nice area where nobody bothers you. Finding the right housing situation is often a hit and miss enterprise. There are some good situations that don't cost much, but finding them isn't easy. A good credit rating which comes from not abusing credit cards, and which is the result of paying all of your bills on time, will help you be accepted into a good apartment.

If you lack income or a credit history, you might consider renting a room in someone's house. However, that individual is likely to do a background check on you. If a lot of people are turning you down for housing, you may want to google yourself and see what shows up.

CONCLUSION: THE LIFE OF A PERSON WITH SCHIZOPHRENIA IS HARD BUT NOT IMPOSSIBLE

One way of approaching life is to set attainable goals and work toward achieving them. This worked for me in my electronics career. Then, later, I had a period of time in which I wasn't processing information well enough to have a career. And then, still later, my processing and my contemplation came back after remaining stabilized on medication for enough years. And that was when I decided I wanted to be a writer. I believed that becoming a writer would be a difficult but not impossible goal, even while other people didn't believe I could do it. A person is not always having delusions of grandeur when attempting to achieve something difficult.

On the other hand, if you were schizophrenic and told me you were going to become an astronaut, or the President, I wouldn't believe you. If you told me that you wanted to become an engineer or a computer programmer, I would hesitate to call you grandiose.

Learning engineering, programming, or doing some kind of professional work like that is difficult but not necessarily impossible for someone with schizophrenia. There are some people who work in high level technical jobs who have a mental illness. You may find that you will have to work harder at it than would a typical person because you will need to deal with various aspects of the disability, while at the same time doing the work. Working at a technical job, for someone with schizophrenia, may be a better work situation than trying to become, say, a short order cook, who must do multitasking and must work very quickly. Also, engineers usually get paid more than cooks do.

Many things are achievable in life if you cooperate with treatment.

Concerning merely enjoying life, that is a separate issue. The approach to enjoying life consists of discarding negative or insistent thoughts and focusing on things in your environment that bring peace and pleasure. The best way to enjoy your life is to focus on what is happening right now, and not to get caught up in emotionally anticipating the future. If you want to enjoy life, I would recommend Zen meditation. Zen meditation, will make you feel better in life, however it won't pay your bills.

I heard someone say that in a person's twenties, they are trying to establish relationships and careers. However, numerous people in their fifties who do not have a mental illness are still struggling with relationships and careers.

Settling for a life of doing nothing but sitting around and smoking cigarettes may not require much effort, but it doesn't yield much. Sitting around and smoking or watching television is certainly a cut above being psychotic and being locked up, however.

Do what you can, and as they say, don't sweat the small stuff. Good luck.